DESIGN SUPPL
ENGINEERING MECHANICS COMBINED

Seventh Edition

Design Supplement Engineering Mechanics Combined

Seventh Edition

Russell C. Hibbeler

PRENTICE HALL Upper Saddle River, NJ 07458

A Simon & Schuster/Viacom Company
Upper Saddle River, NJ 07458

10 9 8 7 6 5 4 3 2 1

ISBN 0-13-569062-5
Printed in the United States of America

Design Supplement Engineering Mechanics Combined

Seventh Edition

[To be assigned after coverage of Section 6.2 or 6.4]

D–1. DESIGN OF A BRIDGE TRUSS

A bridge having a horizontal top cord is to span between the two piers *A* and *B*. It is required that a pin-connected truss be used, consisting of steel members bolted together to steel gusset plates, such as the one shown in the figure. The end supports are assumed to be a pin at *A* and a roller at *B*. A vertical loading of 5 kN is to be supported within the middle 3 m of the span. This load can be applied in part to several joints on the top cord within this region, or to a single joint at the middle of the top cord. The force of the wind and the weight of the members are to be neglected.

Assume the maximum tensile force in each member cannot exceed 4 kN; and regardless of the length of the member, the maximum compressive force cannot exceed 2.5 kN. Design the most economical truss that will support the loading. The members cost $3.50/m, and the gusset plates cost $8.00 each. Submit your cost analysis for the materials, along with the scaled drawing of the truss, identifying on this drawing the tensile and compressive force in each member. Also, include your calculations of the complete force analysis.

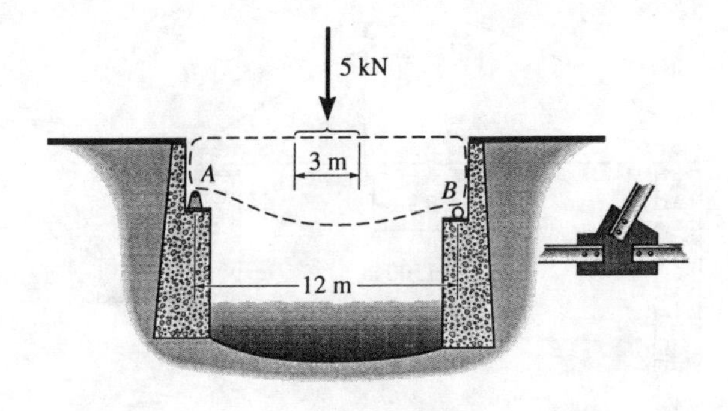

D–2. DESIGN OF A ROOF TRUSS

A roof is to be suspended between two walls. It is required that a pin-connected truss be used, consisting of wood members nailed together with steel gusset plates, such as the one shown in the figure. The end supports are assumed to be a pin at *A* and a roller at *B*. A vertical loading of 1500 lb is to be supported within the middle 15 ft of the span. This load can be applied in part to several joints along the bottom cord within this region, or to a single joint. Neglect the force of wind on the roof and the weight of the members.

The height h of the truss is restricted to be within the range $3 \text{ ft} \leq h \leq 15 \text{ ft}$. Assume the maximum tensile force in each member cannot exceed 750 lb; and regardless of the length of the member, the maximum compressive force cannot exceed 300 lb. Design the most economical truss that will support the loading. The wood members cost \$0.75/ft, and the gusset plates cost \$2.50 each. Submit your cost analysis for the materials, along with a scaled drawing of the truss, identifying on this drawing the tensile and compressive force in each member. Also, include your calculations of the complete force analysis.

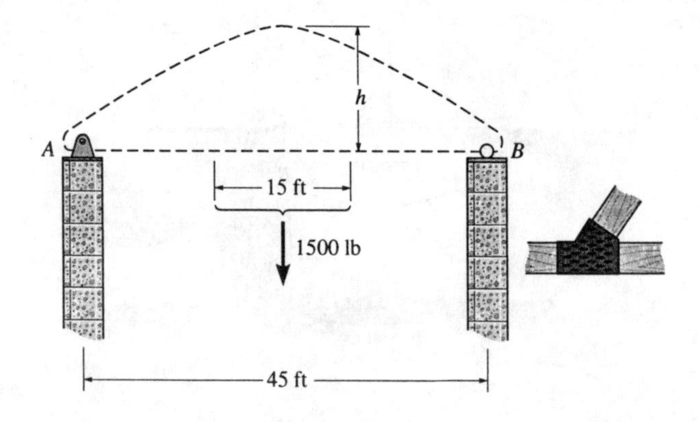

[To be assigned after coverage of Section 6.6]

D–3. DESIGN OF A PULLEY SYSTEM

The steel beam *AB*, having a length of 5 m and a mass of 7 Mg, is to be hoisted in its horizontal position to a height of 4 m. Design a pulley-and-rope system, which can be suspended from the overhead beam *CD*, that will allow a single worker to hoist the beam. Assume that the maximum (comfortable) force that he can apply to the rope is 180 N. Submit a drawing of your design, specify its approximate material cost, and discuss the safety aspects of its operation. Rope costs $1.25/m and each pulley costs $3.00.

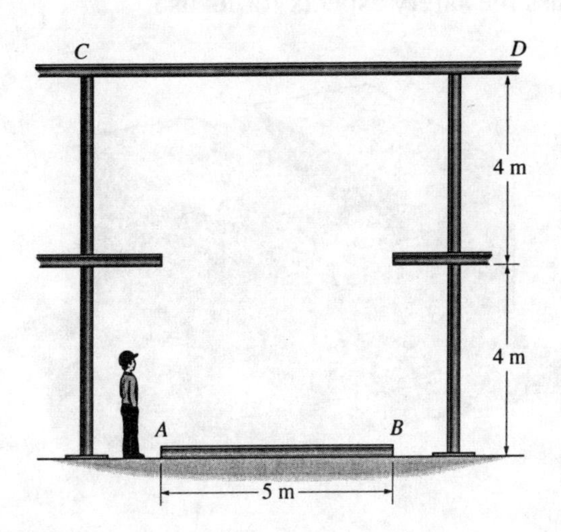

D–4. DESIGN OF A DEVICE TO LIFT CONCRETE BLOCKS

Concrete blocks used for erosion control of bridge and highway embankments are manufactured with a locking groove in them having the dimensions shown. If each block has a weight of 500 lb, design a device that can be used to lift a block by fitting the device down within the groove. (Slipping the device into the ends of the groove is prohibitive, since the blocks are stacked and shipped next to each other.) The device is to be made of a smooth material. Submit a scaled drawing of your device, along with a brief explanation of how it works, based on a force analysis. Also, discuss the safety aspects for its use.

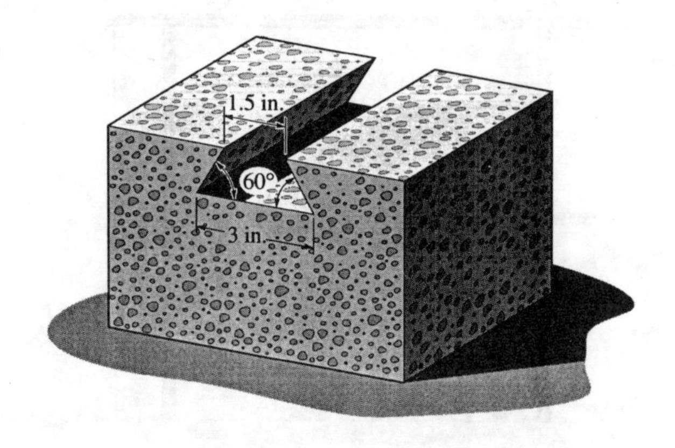

D–5. DESIGN OF A TOOL USED TO POSITION A SUSPENDED LOAD

Heavy loads are suspended from an overhead pulley and each load must be positioned over a depository. Design a tool that can be used to shorten or lengthen the pulley cord *AB* a small amount in order to make the location adjustment. Assume the worker can apply a maximum (comfortable) force of 25 lb to the tool, and the maximum force in cord *AB* is 500 lb. Submit a scaled drawing of the tool, and a brief paragraph to explain how it works using a force analysis. Include a discussion on the safety aspects of its use.

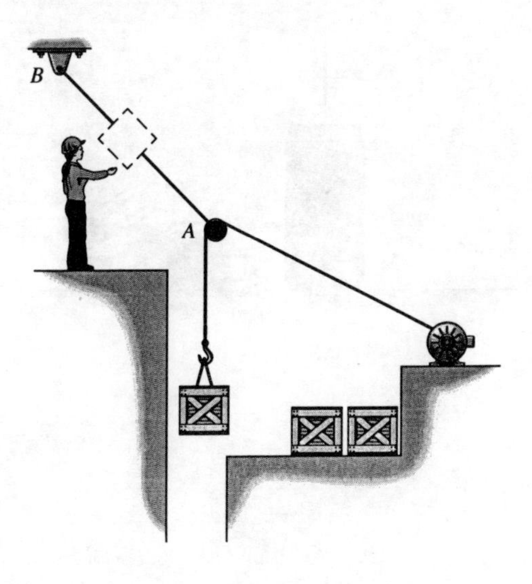

D–6. DESIGN OF A CAN CRUSHER

In an effort to conserve resources a homeowner decides to make a device to crush standard soft-drink cans for recycling. Design a can-crushing device that operates by hand with an applied force of 3 lb, and is easy to construct and operate. Submit a scale drawing of the device, including a force analysis, and discuss the safety and reliability of its use. For this project, perform an experiment to obtain the magnitude of force needed to crush an empty can to a height of 1 in.

D–7. DESIGN OF A FENCE-POST REMOVER

A farmer wishes to remove several fence posts. Each post is buried 18 in. in the ground and will require a maximum vertical pulling force of 175 lb to remove it. He can use his truck to develop the force, but he needs to devise a method for their removal without breaking the posts. Design a method that can be used, considering the only materials available are a strong rope and several pieces of wood having various sizes and lengths. Submit a sketch of your design and discuss the safety and reliability of its use. Also, provide a force analysis to show how it works and why it will cause minimal damage to a post while it is being removed.

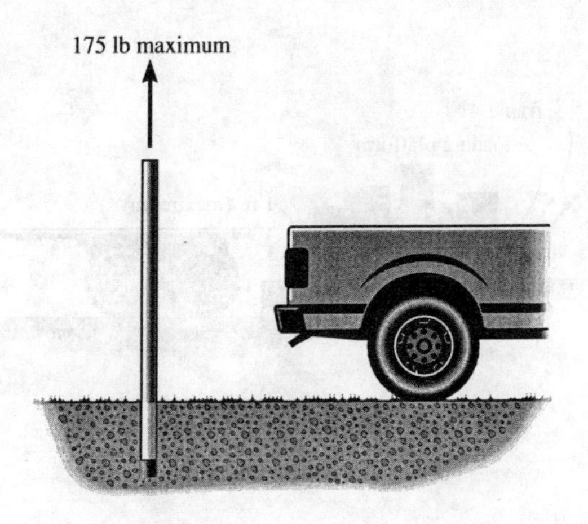

D–8. DESIGN OF A CART LIFT

A hand cart is used to move a load from one loading dock to another. Any dock will have a different elevation relative to the bed of a truck that backs up to it. It is necessary that the loading platform on the hand cart will bring the load resting on it up to the elevation of each truck bed as shown. The maximum elevation difference between the frame of the hand cart and a truck bed is 1 ft. Design a hand-operated mechanical system that will allow the load to be lifted this distance from the frame of the hand cart. Assume the operator can exert a (comfortable) force of 20 lb to make the lift, and that the maximum load, applied to the center of the loading platform, is 400 lb. Submit a scaled drawing of your design, and explain how it works based on a force analysis.

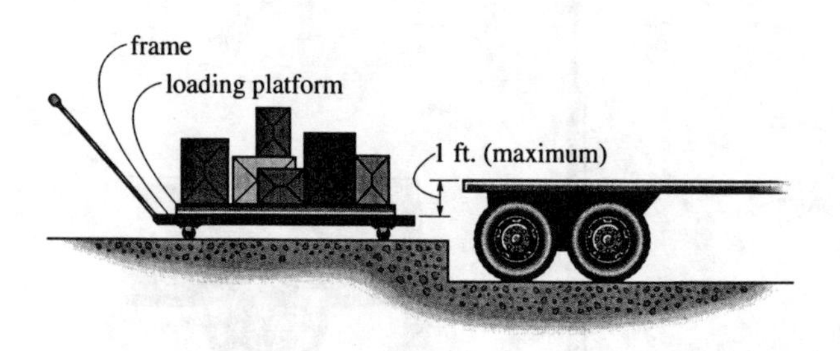

D–9. DESIGN OF A DUMPING DEVICE

Industrial waste is transported in dumpster carts to a centralized garbage bin. Each cart has the dimensions shown, and the maximum weight of a cart and its contents is anticipated to be 500 lb. Design a mechanical means of tipping the cart and allowing its contents to be dumped into the bin. Assume that a worker can exert a comfortable force of 20 lb on any handle used in the dumping operation. Submit a sketch of your design and explain how it works. Include a force analysis and discuss the safety and reliability of its use.

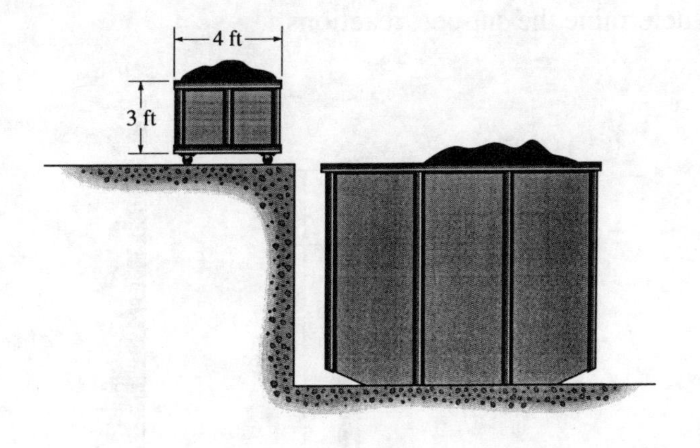

D–10. DESIGN OF A UTILITY POLE SUPPORTING STRUCTURE

An electric utility pole supports wires that exert a tension of 175 lb at the top of the pole in the direction shown. Although the pole is to be somewhat buried at its end *A,* it would be conservative to assume that this connection is a pin. Considering that the pole is located in a restricted space, devise a means of support that will hold the pole in a stable position. The support cannot be attached to the adjacent building or to a point on the sidewalk or roadway. Pedestrian clearance and safety should be considered in the design. Also, the economics of construction and its reliability should be addressed. Submit your design and discuss these issues. Also, provide the calculations used to determine the support reactions.

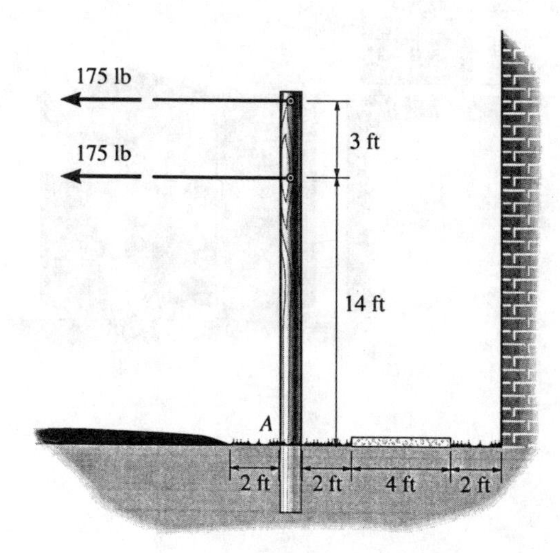

[To be assigned after coverage of Section 8.2]

D–11. DESIGN OF A ROPE-AND-PULLEY SYSTEM FOR PULLING A CRATE UP A RAMP

A large 300-kg packing crate is to be hoisted up the 25° ramp. The coefficient of static friction between the ramp and a crate is $\mu_s = 0.5$, and the coefficient of kinetic friction is $\mu_k = 0.4$. Using a system of ropes and pulleys, design a method that will allow a single worker to pull each crate up the ramp. Pulleys can be attached to any point on the wall *AB*. Assume the worker can exert a maximum (comfortable) pull of 200 N on a rope. Submit a drawing of your design and a force analysis to show how it operates. Estimate the material cost required for its construction. Assume rope costs $0.75/m and a pulley costs $1.80.

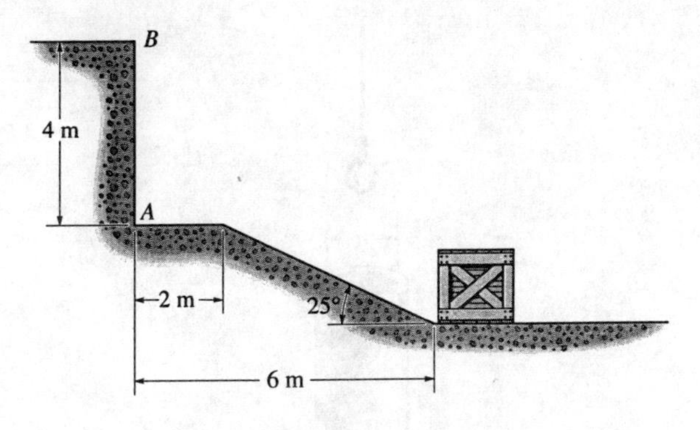

D–12. DESIGN OF A DEVICE FOR LIFTING STAINLESS-STEEL PIPES

Stainless-steel pipes are stacked vertically in a manufacturing plant and are to be moved by an overhead crane from one point to another. The pipes have inner diameters ranging from 100 mm $\leq d \leq$ 250 mm and the maximum mass of any pipe is 500 kg. Design a device that can be connected to the hook and used to lift each pipe. The device should be made of structural steel and should be able to grip the pipe only from its inside surface, since the outside surface is required not to be scratched or damaged. Assume the smallest coefficient of static friction between the two steels is $\mu_s = 0.25$. Submit a scaled drawing of your device, along with a brief explanation of how it works based on a force analysis.

D–13. DESIGN OF A TOOL USED TO RELEASE A SUSPENDED LOAD

A 50-Mg load is suspended from the steel cable, and in an emergency it must be released from the cable so that it falls from the suspended position. Design a steel tool that can be connected to the cable at *A* and used to break the connection. The maximum (comfortable) pull that a worker can exert on the tool is 75 N. If needed, the coefficient of static friction of steel on steel is $\mu_s = 0.30$. Submit a scaled drawing of the tool, and a brief paragraph to explain how it works based on a force analysis. Include a discussion on the safety aspects of its use.

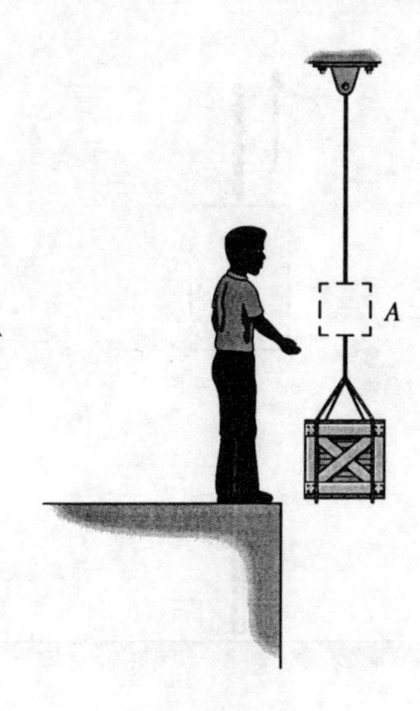

D–14. DESIGN OF A ROD-GRIPPING TOOL

Solid-steel 10-mm-diameter rods each have a mass of 50 kg and are to be individually transported to an assembly line by gripping the rod at one end and suspending it from a chain. Design a steel gripping tool to be used for this purpose, such that it is easy to connect and disconnect and will not damage the end of the rod. Submit a detailed drawing of your device, and include an explanation of how it works based on a force analysis. Discuss the safety aspects of its use. The coefficient of static friction between two steel surfaces is $\mu_s = 0.25$.

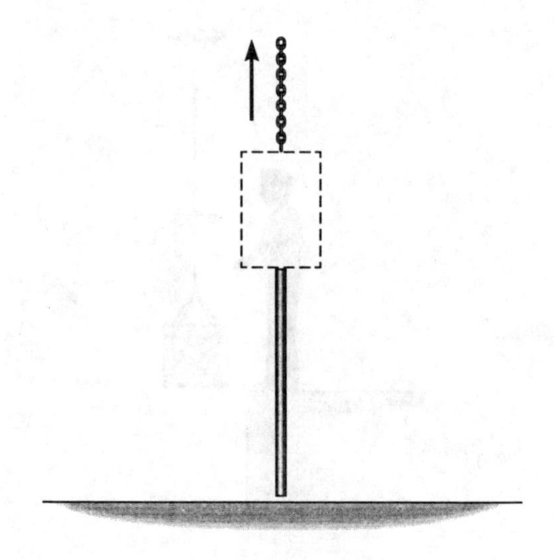

D–15. DESIGN OF A TOOL USED TO TURN PLASTIC PIPE

PVC plastic is often used for sewer pipe. If the outer diameter of any pipe ranges from 4 in. $\leq d \leq$ 8 in., design a tool that can be used by a worker in order to turn the pipe when it is subjected to a maximum anticipated ground resistance of 80 lb·ft. The device is to be made of steel and should be designed so that is does not cut into the pipe and leave any significant marks on its surface. Assume a worker can apply a maximum (comfortable) pull of 40 lb, and take the minimum coefficient of static friction between the PVC and the steel to be $\mu_s = 0.35$. Submit a scaled drawing of the device, and a brief paragraph to explain how it works based on a force analysis.

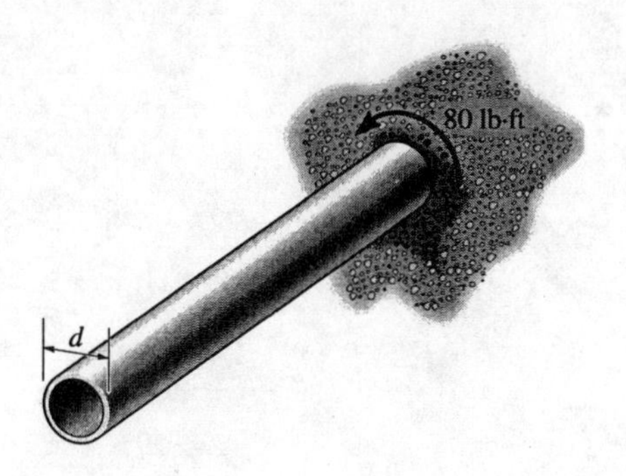

[To be assigned after coverage of Section 12.5]

D–1. DESIGN OF A MARBLE-SORTING DEVICE

Marbles in a manufacturing plant are produced with two diameters, namely, 0.5 in. and 0.75 in. If they both roll off the production chute at 0.5 ft/s, design a device that can be used to sort them out and allow them to fall into separate hoppers. Submit a drawing of the device, and show the path the marbles take and placement of the hoppers relative to the end of the chute.

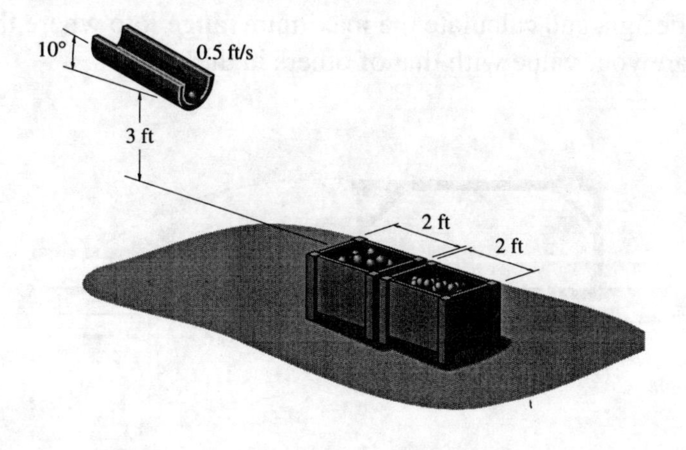

[To be assigned after coverage of Section 13-4]

D–2. DESIGN OF A RAMP CATAPULT

The block B has a mass of 20 kg and is to be catapulted from the table. Design the catapulting mechanism that must be attached to the table and the container of the block, using cables and pulleys. Neglect the mass of the container, assume the operator can exert a constant tension of 120 N on a single cable during operation, and that the maximum movement of his arm is 0.5 m. The coefficient of kinetic friction between the table and container is $\mu_k = 0.2$. Submit a drawing of your design, and calculate the maximum range R to where the block will strike the ground. Compare your value with that of others in the class.

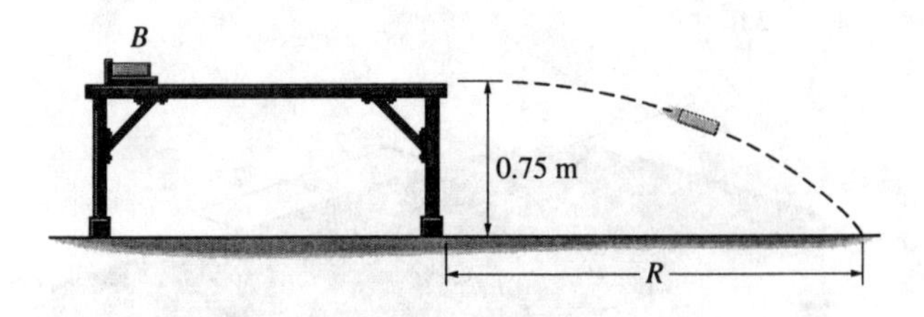

[To be assigned after coverage of Section 14-3]

D–3. DESIGN OF A WATER-BALLOON LAUNCHER

Design a method for launching a 0.25-lb water balloon. Hold a contest with other students to see who can launch the balloon the furthest or hit a target. Materials should consist of a single rubber band of specified length and stiffness, and if necessary no more than three pieces of wood of specified size. Submit a report to show your calculations of where the balloon is predicted to strike the ground from the point at which it was launched. Compare this with the actual value R and discuss why the two distances are different.

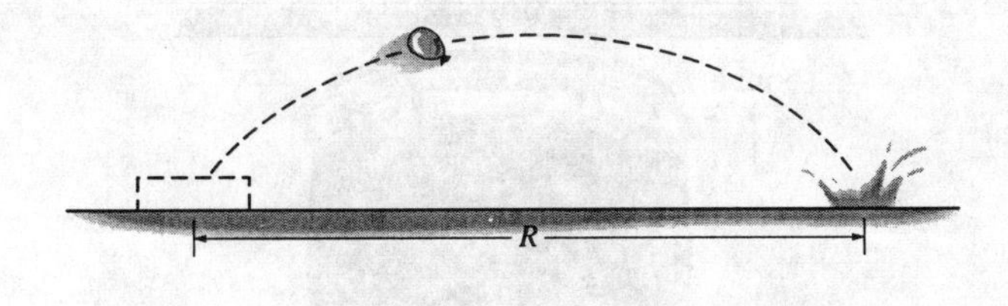

D–4. DESIGN OF A CAR BUMPER

The body of an automobile is to be protected by a spring-loaded bumper, which is attached to the automobile's frame. Design the bumper so that it will stop a 3500-lb car traveling freely at 5 mi/h and not deform the springs more than 3 in. Submit a sketch of your design showing the placement of the springs and their stiffness. Plot the load-deflection diagram for the bumper during a direct collision with a rigid wall, and also plot the deceleration of the car as a function of the springs' displacement.

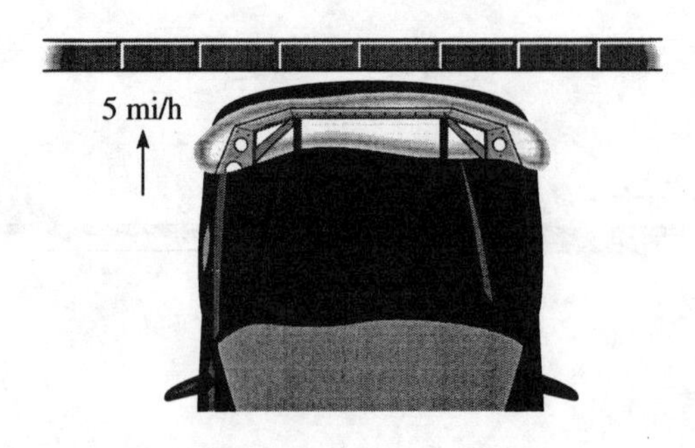

[To be assigned after coverage of Section 14-4]

D–5. DESIGN OF AN ELEVATOR HOIST

It is required that an elevator and its contents, having a maximum weight of 500 lb, be lifted $y = 20$ ft, starting and then stopping after 6 seconds. A single motor and cable-winding drum can be mounted anywhere and used for the operation. During any lift or descent the acceleration should not exceed 10 ft/s^2. Design a cable-and-pulley system for the elevator, and estimate the material cost if the cable is \$1.30/ft and pulleys are \$3.50 each. Submit a drawing of your design, and include plots of the power output required of the motor and the elevator's speed versus the height y traveled.

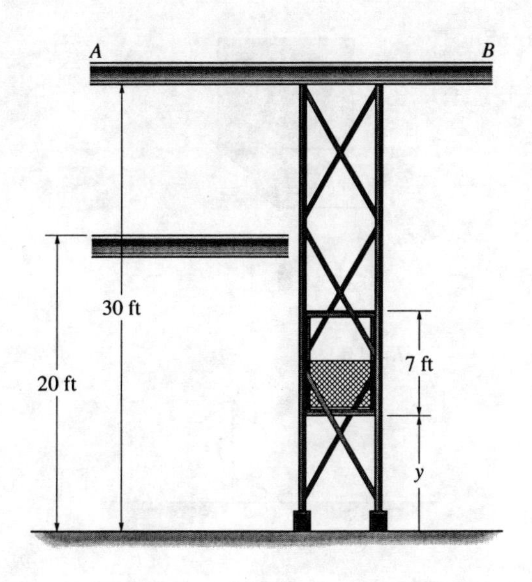

[To be assigned after coverage of Section 14-4]

D–6. DESIGN OF A CRANBERRY SELECTOR

The quality of a cranberry depends upon its firmness, which in turn is related to its bounce. Through experiment, it is found that berries that bounce to a height of $2.5 \leq h' \leq 3.25$ ft, when released from rest at a height of $h = 4$ ft, are appropriate for processing. Using this information, determine the berry's range of the allowable coefficient of restitution, and then design a manner in which good and bad berries can be separated. Submit a drawing of your design, and show calculations as to how the selection and collection of berries is made from your established geometry.

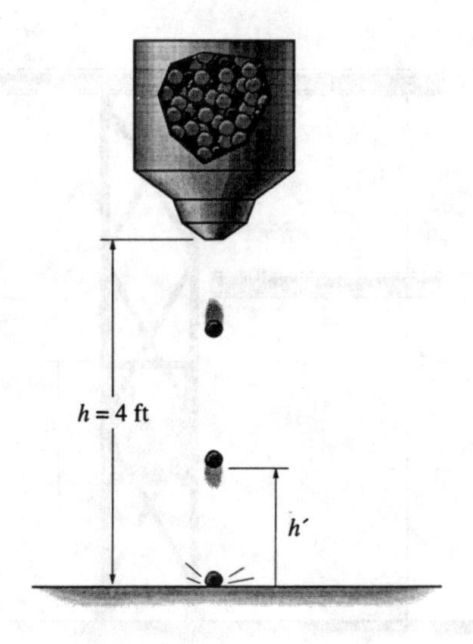

[To be assigned after coverage of Section 16-3]

D–7. DESIGN OF A BELT TRANSMISSION SYSTEM

The wheel A is used in a textile mill and must be rotated counterclockwise at 4 rad/s. This can be done using a motor which is mounted on the platform at the location shown. If the shaft B on the motor can rotate clockwise 50 rad/s, design a method for transmitting the rotation from B to A. Use a series of belts and pulleys as a basis for your design. A belt drives the rotation of wheel A by wrapping around its outer surface and a pulley can be attached to the shaft of the motor. Do not let the length of any belt be longer than 6 ft. Submit a drawing of your design and the calculations of the kinematics. Also, determine the total cost of materials if any belt costs \$2.50, and any pulley costs \$2r, where r is the radius of the pulley.

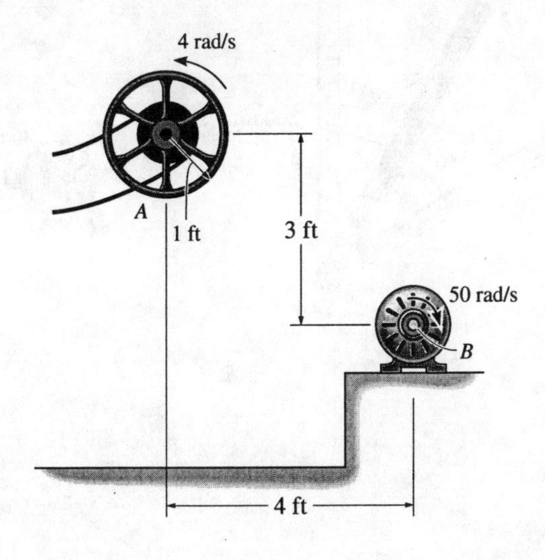

[To be assigned after coverage of Section 16-7]

D–8. DESIGN OF AN OSCILLATING LINK MECHANISM

The operation of a sewing machine requires the 200-mm bar to oscillate back-and-forth through an angle of 60° every 0.2 seconds. A motor having a drive shaft which turns at 40 rad/s is available to provide the necessary power. Specify the location of the motor and design a mechanism required to perform the motion. Submit a drawing of your design, showing the placement of the motor, and compute the velocity and acceleration of the end *A* of the link as a function of its angle of rotation $0° \leq \theta \leq 60°$.

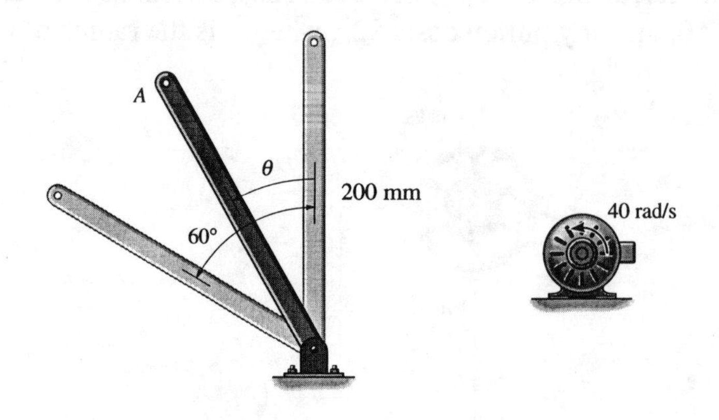

[To be assigned after coverage of Section 16-3]

D–9. DESIGN OF A FEEDING MECHANISM

Boxes are stacked in a bin as shown. It is required that each be ejected onto the rollers every 20 seconds. Design a mechanism that will do this in a reliable manner, using a motor having a rotating shaft of 5 rad/s. Provide a sketch of your design, showing the location of the motor, and plot the velocity and acceleration of each box versus its displacement of 1 ft as it leaves the bin and moves onto the rollers.

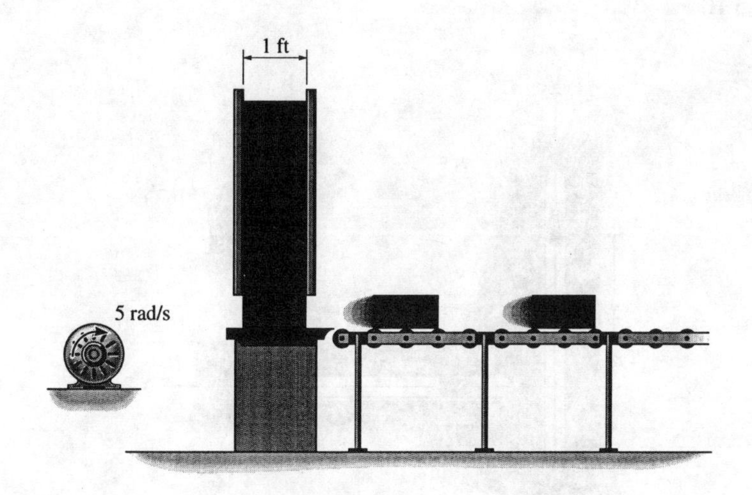

D–10. DESIGN OF A DEVICE FOR CHANGING THE DIRECTION OF LINEAR MOTION

Using a motor which has a rotating shaft of 4 rad/s, design a mechanism that will cause peg *A* to move along the slot on the x axis at a constant rate of 2 ft/s, while simultaneously causing peg *B* to move along the slot on the y axis. Submit a drawing of the mechanism showing the location of the motor. Also, show a plot of the velocity and acceleration of peg *B* versus its position, $0 \leq y \leq 3$ ft.

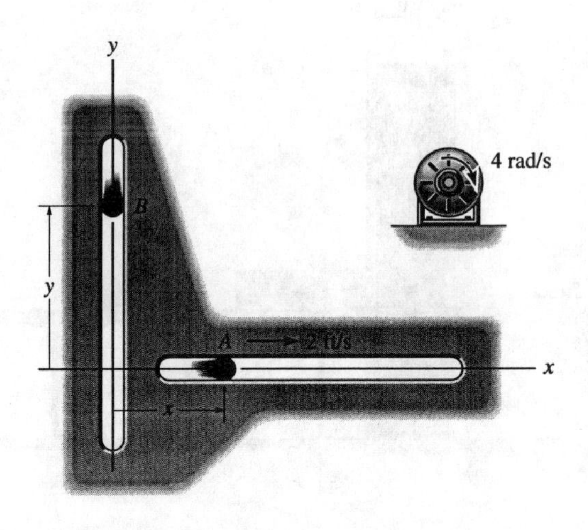

D–11. DESIGN OF A RETRACTABLE AIRCRAFT LANDING-GEAR MECHANISM

The nose wheel of a small plane is attached to member *AB*, which is pinned to the aircraft frame at *B*. Design a mechanism that will allow the wheel to be fully retracted forward; i.e., rotated clockwise 90°, in $t \leq 4$ seconds. Use a hydraulic cylinder which has a closed length of 1.25 ft and if needed a fully extended length of 2 ft. Make sure that your design holds the wheel in a stable position when the wheel is on the ground. Show plots of the angular velocity and angular acceleration of *AB* versus its angular position $0° \leq \theta \leq 90°$.

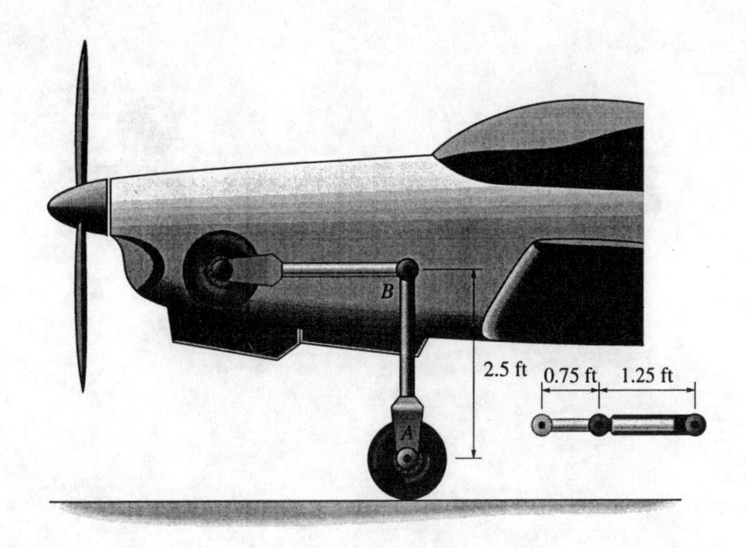

D–12. DESIGN OF A SAW LINK MECHANISM

The saw blade in a lumber mill is required to remain in the horizontal position and undergo a complete back-and-forth motion in 2 seconds. An electric motor, having a shaft rotation of 50 rad/s, is available to power the saw and can be located anywhere. Design a mechanism that will transfer the rotation of the motor's shaft to the saw blade. Submit drawings of your design and calculations of the kinematics of the saw blade. Include a plot of the velocity and acceleration of the saw blade as a function of its horizontal position. Note that to cut through the log the blade must be allowed to move freely downward as well as back and forth.

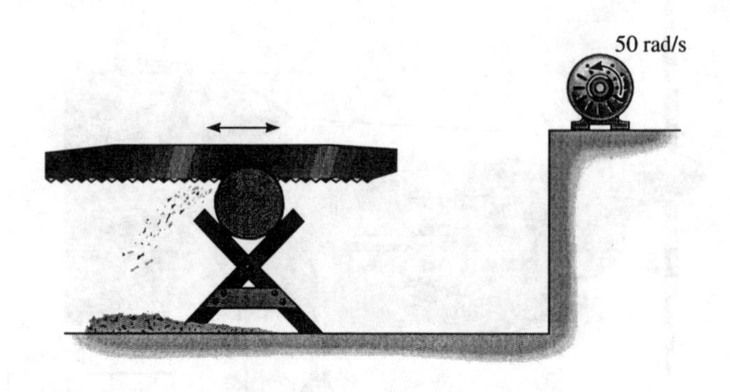

[To be assigned after coverage of Section 17-3]

D–13. DESIGN OF A DYNAMOMETER

In order to test the dynamic strength of cables, an instrument called a *dynamometer* must be used that will measure the tension in a cable when it is used to hoist a very heavy object with accelerated motion. Design such an instrument, based on the use of single or multiple springs so that it can be used on the cable supporting the 300-kg pipe, which is given an upward acceleration of 2 m/s^2. Submit a drawing and explain how your dynamometer operates.

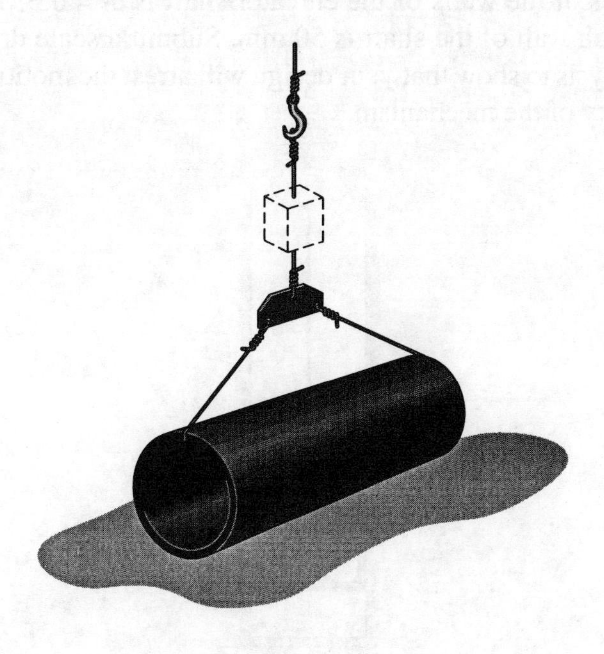

D–14. DESIGN OF A SMALL ELEVATOR BRAKE

A small household elevator is operated using a hoist. For safety purposes it is necessary to install a braking mechanism which will automatically engage in case the cable fails during operation. Design the braking mechanism using steel members and springs. The elevator and its contents have a mass of 300 kg and it travels at 2.5m/s. The maximum allowable deceleration to stop the motion is to be 4 m/s^2. Assume the coefficient of kinetic friction between any steel members and the sides of the walls of the elevator shaft is $\mu_k = 0.3$. The gaps between the elevator frame and each wall of the shaft is 50 mm. Submit a scale drawing of your design along with a force analysis to show that your design will arrest the motion as required. Discuss the safety and reliability of the mechanism.

D–15. SAFETY PERFORMANCE OF A BICYCLE

One of the most common accidents one can have on a bicycle is to flip over the handle bars. Obtain the necessary measurements of a standard-size bicycle and its mass and center of mass. Consider yourself as the rider, with center of mass at your navel. Perform an experiment to determine the coefficient of kinetic friction between the wheels and the pavement. With this data, calculate the possibility of flipping over when (a) only the rear brakes are applied, (b) only the front brakes are applied, and (c) both front and rear brakes are applied simultaneously. What effect does the height of the seat have on these results? Suggest a way to improve the bicycle's design, and write a report on the safety of cycling.